Five of Maxwell's Papers

by James Clerk Maxwell

When observing the spectrum formed by looking at a long vertical slit through a simple prism, I noticed an elongated dark spot running up and down in the blue, and following the motion of the eye as it moved up and down the spectrum, but refusing to pass out of the blue into the other colours. It was plain that the spot belonged both to the eye and to the blue part of the spectrum. The result to which I have come is, that the appearance is due to the yellow spot on the retina, commonly called the Foramen Centrale of Soemmering. The most convenient method of observing the spot is by presenting to the eye in not too rapid succession, blue and yellow glasses, or, still better, allowing blue and yellow papers to revolve slowly before the eye. In this way the spot is seen in the blue. It fades rapidly, but is renewed every time the yellow comes in to relieve the effect of the blue. By using a Nicol's prism along with this apparatus, the brushes of Haidinger are well seen in connexion with the spot, and the fact of the brushes being the spot analysed by polarized light becomes evident. If we look steadily at an object behind a series of bright bars which move in front of it, we shall see a curious bending of the bars as they come up to the place of the yellow spot. The part which comes over the spot seems to start in advance of the rest of the bar, and this would seem to indicate a greater rapidity of sensation at the yellow spot than in the surrounding retina. But I find the experiment difficult, and I hope for better results from more accurate observers.

***

On the Theory of Compound Colours with reference to Mixtures of Blue and Yellow Light.

James Clerk Maxwell

[From the Report of the British Association, 1856.]

When we mix together blue and yellow paint, we obtain green paint. This fact is well known to all who have handled colours; and it is universally admitted that blue and yellow make green. Red, yellow, and blue, being the primary colours among painters, green is regarded as a secondary colour, arising from the mixture of blue and yellow. Newton, however, found that the green of the spectrum was not the same thing as the mixture of two colours of the spectrum, for such a mixture could be separated by the prism, while the green of the spectrum resisted further decomposition. But still it was believed that yellow and blue would make a green, though not that of the spectrum. As far as I am aware, the first experiment on the subject is that of M. Plateau, who, before 1819, made a disc with alternate sectors of prussian blue and gamboge, and observed that, when spinning, the resultant tint was not green, but a neutral gray, inclining sometimes to yellow or blue, but never to green. Prof. J. D. Forbes of Edinburgh made similar experiments in 1849, with the same result. Prof. Helmholtz of Konigsberg, to whom we owe the most complete investigation on visible colour, has given the true explanation of this phenomenon. The result of mixing two coloured powders is not by any means the same as mixing the beams of light which flow from each separately. In the latter case we receive all the light which comes either from the one powder or the other. In the former, much of the light coming from one powder falls on particles of the other, and we receive only that portion which has escaped absorption by one or other. Thus the light coming from a mixture of blue and yellow powder, consists partly of light coming directly from blue particles or yellow particles, and partly of light acted on by both blue and yellow particles. This latter light is green, since the blue stops the red, yellow, and orange, and the yellow stops the blue and violet. I have made experiments on the mixture of blue and yellow light--by rapid rotation, by combined reflexion and transmission, by viewing them out of focus, in stripes, at a great distance, by throwing the colours of the spectrum on a screen, and by receiving them into the eye directly; and I have arranged a portable apparatus by which any one may see the result of this or any other mixture of the colours of the spectrum. In all these cases blue and yellow do not make green. I have also made experiments on the mixture of coloured powders. Those which I used principally were "mineral blue" (from copper) and "chrome-yellow." Other blue and yellow pigments gave curious results, but it was more difficult to make the mixtures, and the greens were less uniform in tint. The mixtures of these colours were made by weight, and were painted on discs of paper, which were afterwards treated in the manner described in my paper "On Colour as perceived by the Eye," in the _Transactions of the Royal Society of Edinburgh_, Vol. XXI. Part 2. The visible effect of the colour is estimated in terms of the standard-coloured papers:-- vermilion (V), ultramarine (U), and emerald-green (E). The accuracy of the results, and their significance, can be best understood by referring to the paper before mentioned. I shall denote mineral blue by B, and chrome-yellow by Y; and B3 Y5 means a mixture of three parts blue and five parts yellow.

Given Colour. Standard Colours. Coefficient V. U. E. of brightness.

B8 , 100 = 2 36 7 .... 45 B7 Y1, 100 = 1 18 17 .... 37 B6 Y2, 100 = 4 11 34 .... 49 B5 Y3, 100 = 9 5 40 .... 54 B4 Y4, 100 = 15 1 40 .... 56 B3 Y5, 100 = 22 - 2 44 .... 64 B2 Y6, 100 = 35 -10 51 .... 76 B1 Y7, 100 = 64 -19 64 .... 109 Y8, 100 = 180 -27 124 277

The columns V, U, E give the proportions of the standard colours which are equivalent to 100 of the given colour; and the sum of V, U, E gives a coefficient, which gives a general idea of the brightness. It will be seen that the first admixture of yellow diminishes the brightness of the blue. The negative values of U indicate that a mixture of V, U, and E cannot be made equivalent to the given colour. The experiments from which these results were taken had the negative values transferred to the other side of the equation. They were all made by means of the colour-top, and were verified by repetition at different times. It may be necessary to remark, in conclusion, with reference to the mode of registering visible colours in terms of three arbitrary standard colours, that it proceeds upon that theory of three primary elements in the sensation of colour, which treats the investigation of the laws of visible colour as a branch of human physiology, incapable of being deduced from the laws of light itself, as set forth in physical optics. It takes advantage of the methods of optics to study vision itself; and its appeal is not to physical principles, but to our consciousness of our own sensations.

*** On an Instrument to illustrate Poinsot's Theory of Rotation.

James Clerk Maxwell

[From the Report of the British Association, 1856.]

In studying the rotation of a solid body according to Poinsot's method, we have to consider the successive positions of the instantaneous axis of rotation with reference both to directions fixed in space and axes assumed in the moving body. The paths traced out by the pole of this axis on the invariable plane and on the _central ellipsoid_ form interesting subjects of mathematical investigation. But when we attempt to follow with our eye the motion of a rotating body, we find it difficult to determine through what point of the body the instantaneous axis passes at any time,--and to determine its path must be still more difficult. I have endeavoured to render visible the path of the instantaneous axis, and to vary the circumstances of motion, by means of a top of the same kind as that used by Mr Elliot, to illustrate precession*. The body of the instrument is a hollow cone of wood, rising from a ring, 7 inches in diameter and 1 inch thick. An iron axis, 8 inches long, screws into the vertex of the cone. The lower extremity has a point of hard steel, which rests in an agate cup, and forms the support of the instrument. An iron nut, three ounces in weight, is made to screw on the axis, and to be fixed at any point; and in the wooden ring are screwed four bolts, of three ounces, working horizontally, and four bolts, of one ounce, working vertically. On the upper part of the axis is placed a disc of card, on which are drawn four concentric rings. Each ring is divided into four quadrants, which are coloured red, yellow, green, and blue. The spaces between the rings are white. When the top is in motion, it is easy to see in which quadrant the instantaneous axis is at any moment and the distance between it and the axis of the instrument; and we observe,--1st. That the instantaneous axis travels in a closed curve, and returns to its original position in the body. 2ndly. That by working the vertical bolts, we can make the axis of the instrument the centre of this closed curve. It will then be one of the principal axes of inertia. 3rdly. That, by working the nut on the axis, we can make the order of colours either red, yellow, green, blue, or the reverse. When the order of colours is in the same direction as the rotation, it indicates that the axis of the instrument is that of greatest moment of inertia. 4thly. That if we screw the two pairs of opposite horizontal bolts to different distances from the axis, the path of the instantaneous pole will no longer be equidistant from the axis, but will describe an ellipse, whose longer axis is in the direction of the mean axis of the instrument. 5thly. That if we now make one of the two horizontal axes less and the other greater than the vertical axis, the instantaneous pole will separate from the axis of the instrument, and the axis will incline more and more till the spinning can no longer go on, on account of the obliquity. It is easy to see that, by attending to the laws of motion, we may produce any of the above effects at pleasure, and illustrate many different propositions by means of the same instrument.

* Transactions of the Royal Scottish Society of Arts, 1855.

*** Address to the Mathematical and Physical Sections of the British Association.

James Clerk Maxwell

[From the British Association Report, Vol. XL.]

[Liverpool, September 15, 1870.]

At several of the recent Meetings of the British Association the varied and important business of the Mathematical and Physical Section has been introduced by an Address, the subject of which has been left to the selection of the President for the time being. The perplexing duty of choosing a subject has not, however, fallen to me.

Professor Sylvester, the President of Section A at the Exeter Meeting, gave us a noble vindication of pure mathematics by laying bare, as it were, the very working of the mathematical mind, and setting before us, not the array of symbols and brackets which form the armoury of the mathematician, or the dry results which are only the monuments of his conquests, but the mathematician himself, with all his human faculties directed by his professional sagacity to the pursuit, apprehension, and exhibition of that ideal harmony which he feels to be the root of all knowledge, the fountain of all pleasure, and the condition of all action. The mathematician has, above all things, an eye for symmetry; and Professor Sylvester has not only recognized the symmetry formed by the combination of his own subject with those of the former Presidents, but has pointed out the duties of his successor in the following characteristic note:--

"Mr Spottiswoode favoured the Section, in his opening Address, with a combined history of the progress of Mathematics and Physics; Dr. Tyndall's address was virtually on the limits of Physical Philosophy; the one here in print," says Prof. Sylvester, "is an attempted faint adumbration of the nature of Mathematical Science in the abstract. What is wanting (like a fourth sphere resting on three others in contact) to build up the Ideal Pyramid is a discourse on the Relation of the two branches (Mathematics and Physics) to, their action and reaction upon, one another, a magnificent theme, with which it is to be hoped that some future President of Section A will crown the edifice and make the Tetralogy (symbolizable by A+A', A, A', AA') complete."

The theme thus distinctly laid down for his successor by our late President is indeed a magnificent one, far too magnificent for any efforts of mine to realize. I have endeavoured to follow Mr Spottiswoode, as with far-reaching vision he distinguishes the systems of science into which phenomena, our knowledge of which is still in the nebulous stage, are growing. I have been carried by the penetrating insight and forcible expression of Dr Tyndall into that sanctuary of minuteness and of power where molecules obey the laws of their existence, clash together in fierce collision, or grapple in yet more fierce embrace, building up in secret the forms of visible things. I have been guided by Prof. Sylvester towards those serene heights

"Where never creeps a cloud, or moves a wind, Nor ever falls the least white star of snow, Nor ever lowest roll of thunder moans, Nor sound of human sorrow mounts to mar Their sacred everlasting calm."

But who will lead me into that still more hidden and dimmer region where Thought weds Fact, where the mental operation of the mathematician and the physical action of the molecules are seen in their true relation? Does not the way to it pass through the very den of the metaphysician, strewed with the remains of former explorers, and abhorred by every man of science? It would indeed be a foolhardy adventure for me to take up the valuable time of the Section by leading you into those speculations which require, as we know, thousands of years even to shape themselves intelligibly.

But we are met as cultivators of mathematics and physics. In our daily work we are led up to questions the same in kind with those of metaphysics; and we approach them, not trusting to the native penetrating power of our own minds, but trained by a long-continued adjustment of our modes of thought to the facts of external nature.

As mathematicians, we perform certain mental operations on the symbols of number or of quantity, and, by proceeding step by step from more simple to more complex operations, we are enabled to express the same thing in many different forms. The equivalence of these different forms, though a necessary consequence of self-evident axioms, is not always, to our minds, self-evident; but the mathematician, who by long practice has acquired a familiarity with many of these forms, and has become expert in the processes which lead from one to another, can often transform a perplexing expression into another which explains its meaning in more intelligible language.

As students of Physics we observe phenomena under varied circumstances, and endeavour to deduce the laws of their relations. Every natural phenomenon is, to our minds, the result of an infinitely complex system of conditions. What we set ourselves to do is to unravel these conditions, and by viewing the phenomenon in a way which is in itself partial and imperfect, to piece out its features one by one, beginning with that which strikes us first, and thus gradually learning how to look at the whole phenomenon so as to obtain a continually greater degree of clearness and distinctness. In this process, the feature which presents itself most forcibly to the untrained inquirer may not be that which is considered most fundamental by the experienced man of science; for the success of any physical investigation depends on the judicious selection of what is to be observed as of primary importance, combined with a voluntary abstraction of the mind from those features which, however attractive they appear, we are not yet sufficiently advanced in science to investigate with profit.

Intellectual processes of this kind have been going on since the first formation of language, and are going on still. No doubt the feature which strikes us first and most forcibly in any phenomenon, is the pleasure or the pain which accompanies it, and the agreeable or disagreeable results which follow after it. A theory of nature from this point of view is embodied in many of our words and phrases, and is by no means extinct even in our deliberate opinions.

It was a great step in science when men became convinced that, in order to understand the nature of things, they must begin by asking, not whether a thing is good or bad, noxious or beneficial, but of what kind is it? and how much is there of it? Quality and Quantity were then first recognized as the primary features to be observed in scientific inquiry.

As science has been developed, the domain of quantity has everywhere encroached on that of quality, till the process of scientific inquiry seems to have become simply the measurement and registration of quantities, combined with a mathematical discussion of the numbers thus obtained. It is this scientific method of directing our attention to those features of phenomena which may be regarded as quantities which brings physical research under the influence of mathematical reasoning. In the work of the Section we shall have abundant examples of the successful application of this method to the most recent conquests of science; but I wish at present to direct your attention to some of the reciprocal effects of the progress of science on those elementary conceptions which are sometimes thought to be beyond the reach of change.

If the skill of the mathematician has enabled the experimentalist to see that the quantities which he has measured are connected by necessary relations, the discoveries of physics have revealed to the mathematician new forms of quantities which he could never have imagined for himself.

Of the methods by which the mathematician may make his labours most useful to the student of nature, that which I think is at present most important is the systematic classification of quantities.

The quantities which we study in mathematics and physics may be classified in two different ways.

The student who wishes to master any particular science must make himself familiar with the various kinds of quantities which belong to that science. When he understands all the relations between these quantities, he regards them as forming a connected system, and he classes the whole system of quantities together as belonging to that particular science. This classification is the most natural from a physical point of view, and it is generally the first in order of time.

But when the student has become acquainted with several different sciences, he finds that the mathematical processes and trains of reasoning in one science resemble those in another so much that his knowledge of the one science may be made a most useful help in the study of the other.

When he examines into the reason of this, he finds that in the two sciences he has been dealing with systems of quantities, in which the mathematical forms of the relations of the quantities are the same in both systems, though the physical nature of the quantities may be utterly different.

He is thus led to recognize a classification of quantities on a new principle, according to which the physical nature of the quantity is subordinated to its mathematical form. This is the point of view which is characteristic of the mathematician; but it stands second to the physical aspect in order of time, because the human mind, in order to conceive of different kinds of quantities, must have them presented to it by nature.

I do not here refer to the fact that all quantities, as such, are subject to the rules of arithmetic and algebra, and are therefore capable of being submitted to those dry calculations which represent, to so many minds, their only idea of mathematics.

The human mind is seldom satisfied, and is certainly never exercising its highest functions, when it is doing the work of a calculating machine. What the man of science, whether he is a mathematician or a physical inquirer, aims at is, to acquire and develope clear ideas of the things he deals with. For this purpose he is willing to enter on long calculations, and to be for a season a calculating machine, if he can only at last make his ideas clearer.

But if he finds that clear ideas are not to be obtained by means of processes the steps of which he is sure to forget before he has reached the conclusion, it is much better that he should turn to another method, and try to understand the subject by means of well-chosen illustrations derived from subjects with which he is more familiar.

We all know how much more popular the illustrative method of exposition is found, than that in which bare processes of reasoning and calculation form the principal subject of discourse.

Now a truly scientific illustration is a method to enable the mind to grasp some conception or law in one branch of science, by placing before it a conception or a law in a different branch of science, and directing the mind to lay hold of that mathematical form which is common to the corresponding ideas in the two sciences, leaving out of account for the present the difference between the physical nature of the real phenomena.

The correctness of such an illustration depends on whether the two systems of ideas which are compared together are really analogous in form, or whether, in other words, the corresponding physical quantities really belong to the same mathematical class. When this condition is fulfilled, the illustration is not only convenient for teaching science in a pleasant and easy manner, but the recognition of the formal analogy between the two systems of ideas leads to a knowledge of both, more profound than could be obtained by studying each system separately.

There are men who, when any relation or law, however complex, is put before them in a symbolical form, can grasp its full meaning as a relation among abstract quantities. Such men sometimes treat with indifference the further statement that quantities actually exist in nature which fulfil this relation. The mental image of the concrete reality seems rather to disturb than to assist their contemplations. But the great majority of mankind are utterly unable, without long training, to retain in their minds the unembodied symbols of the pure mathematician, so that, if science is ever to become popular, and yet remain scientific, it must be by a profound study and a copious application of those principles of the mathematical classification of quantities which, as we have seen, lie at the root of every truly scientific illustration.

There are, as I have said, some minds which can go on contemplating with satisfaction pure quantities presented to the eye by symbols, and to the mind in a form which none but mathematicians can conceive.

There are others who feel more enjoyment in following geometrical forms, which they draw on paper, or build up in the empty space before them.

Others, again, are not content unless they can project their whole physical energies into the scene which they conjure up. They learn at what a rate the planets rush through space, and they experience a delightful feeling of exhilaration. They calculate the forces with which the heavenly bodies pull at one another, and they feel their own muscles straining with the effort.

To such men momentum, energy, mass are not mere abstract expressions of the results of scientific inquiry. They are words of power, which stir their souls like the memories of childhood.

For the sake of persons of these different types, scientific truth should be presented in different forms, and should be regarded as equally scientific whether it appears in the robust form and the vivid colouring of a physical illustration, or in the tenuity and paleness of a symbolical expression.

Time would fail me if I were to attempt to illustrate by examples the scientific value of the classification of quantities. I shall only mention the name of that important class of magnitudes having direction in space which Hamilton has called vectors, and which form the subject-matter of the Calculus of Quaternions, a branch of mathematics which, when it shall have been thoroughly understood by men of the illustrative type, and clothed by them with physical imagery, will become, perhaps under some new name, a most powerful method of communicating truly scientific knowledge to persons apparently devoid of the calculating spirit.

The mutual action and reaction between the different departments of human thought is so interesting to the student of scientific progress, that, at the risk of still further encroaching on the valuable time of the Section, I shall say a few words on a branch of physics which not very long ago would have been considered rather a branch of metaphysics. I mean the atomic theory, or, as it is now called, the molecular theory of the constitution of bodies.

Not many years ago if we had been asked in what regions of physical science the advance of discovery was least apparent, we should have pointed to the hopelessly distant fixed stars on the one hand, and to the inscrutable delicacy of the texture of material bodies on the other.

Indeed, if we are to regard Comte as in any degree representing the scientific opinion of his time, the research into what takes place beyond our own solar system seemed then to be exceedingly unpromising, if not altogether illusory.

The opinion that the bodies which we see and handle, which we can set in motion or leave at rest, which we can break in pieces and destroy, are composed of smaller bodies which we cannot see or handle, which are always in motion, and which can neither be stopped nor broken in pieces, nor in any way destroyed or deprived of the least of their properties, was known by the name of the Atomic theory. It was associated with the names of Democritus, Epicurus, and Lucretius, and was commonly supposed to admit the existence only of atoms and void, to the exclusion of any other basis of things from the universe.

In many physical reasonings and mathematical calculations we are accustomed to argue as if such substances as air, water, or metal, which appear to our senses uniform and continuous, were strictly and mathematically uniform and continuous.

We know that we can divide a pint of water into many millions of portions, each of which is as fully endowed with all the properties of water as the whole pint was; and it seems only natural to conclude that we might go on subdividing the water for ever, just as we can never come to a limit in subdividing the space in which it is contained. We have heard how Faraday divided a grain of gold into an inconceivable number of separate particles, and we may see Dr Tyndall produce from a mere suspicion of nitrite of butyle an immense cloud, the minute visible portion of which is still cloud, and therefore must contain many molecules of nitrite of butyle.

But evidence from different and independent sources is now crowding in upon us which compels us to admit that if we could push the process of subdivision still further we should come to a limit, because each portion would then contain only one molecule, an individual body, one and indivisible, unalterable by any power in nature.

Even in our ordinary experiments on very finely divided matter we find that the substance is beginning to lose the properties which it exhibits when in a large mass, and that effects depending on the individual action of molecules are beginning to become prominent.

The study of these phenomena is at present the path which leads to the development of molecular science.

That superficial tension of liquids which is called capillary attraction is one of these phenomena. Another important class of phenomena are those which are due to that motion of agitation by which the molecules of a liquid or gas are continually working their way from one place to another, and continually changing their course, like people hustled in a crowd.

On this depends the rate of diffusion of gases and liquids through each other, to the study of which, as one of the keys of molecular science, that unwearied inquirer into nature's secrets, the late Prof. Graham, devoted such arduous labour.

The rate of electrolytic conduction is, according to Wiedemann's theory, influenced by the same cause; and the conduction of heat in fluids depends probably on the same kind of action. In the case of gases, a molecular theory has been developed by Clausius and others, capable of mathematical treatment, and subjected to experimental investigation; and by this theory nearly every known mechanical property of gases has been explained on dynamical principles; so that the properties of individual gaseous molecules are in a fair way to become objects of scientific research.

Now Mr Stoney has pointed out[1] that the numerical results of experiments on gases render it probable that the mean distance of their particles at the ordinary temperature and pressure is a quantity of the same order of magnitude as a millionth of a millimetre, and Sir William Thomson has since[2] shewn, by several independent lines of argument, drawn from phenomena so different in themselves as the electrification of metals by contact, the tension of soap-bubbles, and the friction of air, that in ordinary solids and liquids the average distance between contiguous molecules is less than the hundred-millionth, and greater than the two-thousand-millionth of a centimetre.

[1] Phil. Mag., Aug. 1868. [2] Nature, March 31, 1870.

These, of course, are exceedingly rough estimates, for they are derived from measurements some of which are still confessedly very rough; but if at the present time, we can form even a rough plan for arriving at results of this kind, we may hope that, as our means of experimental inquiry become more accurate and more varied, our conception of a molecule will become more definite, so that we may be able at no distant period to estimate its weight with a greater degree of precision.

A theory, which Sir W. Thomson has founded on Helmholtz's splendid hydrodynamical theorems, seeks for the properties of molecules in the ring vortices of a uniform, frictionless, incompressible fluid. Such whirling rings may be seen when an experienced smoker sends out a dexterous puff of smoke into the still air, but a more evanescent phenomenon it is difficult to conceive. This evanescence is owing to the viscosity of the air; but Helmholtz has shewn that in a perfect fluid such a whirling ring, if once generated, would go on whirling for ever, would always consist of the very same portion of the fluid which was first set whirling, and could never be cut in two by any natural cause. The generation of a ring-vortex is of course equally beyond the power of natural causes, but once generated, it has the properties of individuality, permanence in quantity, and indestructibility. It is also the recipient of impulse and of energy, which is all we can affirm of matter; and these ring-vortices are capable of such varied connexions and knotted self-involutions, that the properties of differently knotted vortices must be as different as those of different kinds of molecules can be.

If a theory of this kind should be found, after conquering the enormous mathematical difficulties of the subject, to represent in any degree the actual properties of molecules, it will stand in a very different scientific position from those theories of molecular action which are formed by investing the molecule with an arbitrary system of central forces invented expressly to account for the observed phenomena.

In the vortex theory we have nothing arbitrary, no central forces or occult properties of any other kind. We have nothing but matter and motion, and when the vortex is once started its properties are all determined from the original impetus, and no further assumptions are possible.

Even in the present undeveloped state of the theory, the contemplation of the individuality and indestructibility of a ring-vortex in a perfect fluid cannot fail to disturb the commonly received opinion that a molecule, in order to be permanent, must be a very hard body.

In fact one of the first conditions which a molecule must fulfil is, apparently, inconsistent with its being a single hard body. We know from those spectroscopic researches which have thrown so much light on different branches of science, that a molecule can be set into a state of internal vibration, in which it gives off to the surrounding medium light of definite refrangibility--light, that is, of definite wave-length and definite period of vibration. The fact that all the molecules (say, of hydrogen) which we can procure for our experiments, when agitated by heat or by the passage of an electric spark, vibrate precisely in the same periodic time, or, to speak more accurately, that their vibrations are composed of a system of simple vibrations having always the same periods, is a very remarkable fact.

I must leave it to others to describe the progress of that splendid series of spectroscopic discoveries by which the chemistry of the heavenly bodies has been brought within the range of human inquiry. I wish rather to direct your attention to the fact that, not only has every molecule of terrestrial hydrogen the same system of periods of free vibration, but that the spectroscopic examination of the light of the sun and stars shews that, in regions the distance of which we can only feebly imagine, there are molecules vibrating in as exact unison with the molecules of terrestrial hydrogen as two tuning-forks tuned to concert pitch, or two watches regulated to solar time.

Now this absolute equality in the magnitude of quantities, occurring in all parts of the universe, is worth our consideration.

The dimensions of individual natural bodies are either quite indeterminate, as in the case of planets, stones, trees, &c., or they vary within moderate limits, as in the case of seeds, eggs, &c.; but even in these cases small quantitative differences are met with which do not interfere with the essential properties of the body.

Even crystals, which are so definite in geometrical form, are variable with respect to their absolute dimensions.

Among the works of man we sometimes find a certain degree of uniformity.

There is a uniformity among the different bullets which are cast in the same mould, and the different copies of a book printed from the same type.

If we examine the coins, or the weights and measures, of a civilized country, we find a uniformity, which is produced by careful adjustment to standards made and provided by the state. The degree of uniformity of these national standards is a measure of that spirit of justice in the nation which has enacted laws to regulate them and appointed officers to test them.

This subject is one in which we, as a scientific body, take a warm interest; and you are all aware of the vast amount of scientific work which has been expended, and profitably expended, in providing weights and measures for commercial and scientific purposes.

The earth has been measured as a basis for a permanent standard of length, and every property of metals has been investigated to guard against any alteration of the material standards when made. To weigh or measure any thing with modern accuracy, requires a course of experiment and calculation in which almost every branch of physics and mathematics is brought into requisition.

Yet, after all, the dimensions of our earth and its time of rotation, though, relatively to our present means of comparison, very permanent, are not so by any physical necessity. The earth might contract by cooling, or it might be enlarged by a layer of meteorites falling on it, or its rate of revolution might slowly slacken, and yet it would continue to be as much a planet as before.

But a molecule, say of hydrogen, if either its mass or its time of vibration were to be altered in the least, would no longer be a molecule of hydrogen.

If, then, we wish to obtain standards of length, time, and mass which shall be absolutely permanent, we must seek them not in the dimensions, or the motion, or the mass of our planet, but in the wave-length, the period of vibration, and the absolute mass of these imperishable and unalterable and perfectly similar molecules.

When we find that here, and in the starry heavens, there are innumerable multitudes of little bodies of exactly the same mass, so many, and no more, to the grain, and vibrating in exactly the same time, so many times, and no more, in a second, and when we reflect that no power in nature can now alter in the least either the mass or the period of any one of them, we seem to have advanced along the path of natural knowledge to one of those points at which we must accept the guidance of that faith by which we understand that "that which is seen was not made of things which do appear."

One of the most remarkable results of the progress of molecular science is the light it has thrown on the nature of irreversible processes--processes, that is, which always tend towards and never away from a certain limiting state. Thus, if two gases be put into the same vessel, they become mixed, and the mixture tends continually to become more uniform. If two unequally heated portions of the same gas are put into the vessel, something of the kind takes place, and the whole tends to become of the same temperature. If two unequally heated solid bodies be placed in contact, a continual approximation of both to an intermediate temperature takes place.

In the case of the two gases, a separation may be effected by chemical means; but in the other two cases the former state of things cannot be restored by any natural process.

In the case of the conduction or diffusion of heat the process is not only irreversible, but it involves the irreversible diminution of that part of the whole stock of thermal energy which is capable of being converted into mechanical work.

This is Thomson's theory of the irreversible dissipation of energy, and it is equivalent to the doctrine of Clausius concerning the growth of what he calls Entropy.

The irreversible character of this process is strikingly embodied in Fourier's theory of the conduction of heat, where the formulae themselves indicate, for all positive values of the time, a possible solution which continually tends to the form of a uniform diffusion of heat.

But if we attempt to ascend the stream of time by giving to its symbol continually diminishing values, we are led up to a state of things in which the formula has what is called a critical value; and if we inquire into the state of things the instant before, we find that the formula becomes absurd.

We thus arrive at the conception of a state of things which cannot be conceived as the physical result of a previous state of things, and we find that this critical condition actually existed at an epoch not in the utmost depths of a past eternity, but separated from the present time by a finite interval.

This idea of a beginning is one which the physical researches of recent times have brought home to us, more than any observer of the course of scientific thought in former times would have had reason to expect.

But the mind of man is not, like Fourier's heated body, continually settling down into an ultimate state of quiet uniformity, the character of which we can already predict; it is rather like a tree, shooting out branches which adapt themselves to the new aspects of the sky towards which they climb, and roots which contort themselves among the strange strata of the earth into which they delve. To us who breathe only the spirit of our own age, and know only the characteristics of contemporary thought, it is as impossible to predict the general tone of the science of the future as it is to anticipate the particular discoveries which it will make.

Physical research is continually revealing to us new features of natural processes, and we are thus compelled to search for new forms of thought appropriate to these features. Hence the importance of a careful study of those relations between mathematics and Physics which determine the conditions under which the ideas derived from one department of physics may be safely used in forming ideas to be employed in a new department.

The figure of speech or of thought by which we transfer the language and ideas of a familiar science to one with which we are less acquainted may be called Scientific Metaphor.

Thus the words Velocity, Momentum, Force, &c. have acquired certain precise meanings in Elementary Dynamics. They are also employed in the Dynamics of a Connected System in a sense which, though perfectly analogous to the elementary sense, is wider and more general.

These generalized forms of elementary ideas may be called metaphorical terms in the sense in which every abstract term is metaphorical. The characteristic of a truly scientific system of metaphors is that each term in its metaphorical use retains all the formal relations to the other terms of the system which it had in its original use. The method is then truly scientific-- that is, not only a legitimate product of science, but capable of generating science in its turn.

There are certain electrical phenomena, again, which are connected together by relations of the same form as those which connect dynamical phenomena. To apply to these the phrases of dynamics with proper distinctions and provisional reservations is an example of a metaphor of a bolder kind; but it is a legitimate metaphor if it conveys a true idea of the electrical relations to those who have been already trained in dynamics.

Suppose, then, that we have successfully introduced certain ideas belonging to an elementary science by applying them metaphorically to some new class of phenomena. It becomes an important philosophical question to determine in what degree the applicability of the old ideas to the new subject may be taken as evidence that the new phenomena are physically similar to the old.

The best instances for the determination of this question are those in which two different explanations have been given of the same thing.

The most celebrated case of this kind is that of the corpuscular and the undulatory theories of light. Up to a certain point the phenomena of light are equally well explained by both; beyond this point, one of them fails.

To understand the true relation of these theories in that part of the field where they seem equally applicable we must look at them in the light which Hamilton has thrown upon them by his discovery that to every brachistochrone problem there corresponds a problem of free motion, involving different velocities and times, but resulting in the same geometrical path. Professor Tait has written a very interesting paper on this subject.

According to a theory of electricity which is making great progress in Germany, two electrical particles act on one another directly at a distance, but with a force which, according to Weber, depends on their relative velocity, and according to a theory hinted at by Gauss, and developed by Riemann, Lorenz, and Neumann, acts not instantaneously, but after a time depending on the distance. The power with which this theory, in the hands of these eminent men, explains every kind of electrical phenomena must be studied in order to be appreciated.

  Another theory of electricity, which I prefer, denies action at a distance and attributes electric action to tensions and pressures in an all-pervading medium, these stresses being the same in kind with those familiar to engineers, and the medium being identical with that in which light is supposed to be propagated.

  Both these theories are found to explain not only the phenomena by the aid of which they were originally constructed, but other phenomena, which were not thought of or perhaps not known at the time; and both have independently arrived at the same numerical result, which gives the absolute velocity of light in terms of electrical quantities.

  That theories apparently so fundamentally opposed should have so large a field of truth common to both is a fact the philosophical importance of which we cannot fully appreciate till we have reached a scientific altitude from which the true relation between hypotheses so different can be seen.

  I shall only make one more remark on the relation between Mathematics and Physics. In themselves, one is an operation of the mind, the other is a dance of molecules. The molecules have laws of their own, some of which we select as most intelligible to us and most amenable to our calculation. We form a theory from these partial data, and we ascribe any deviation of the actual phenomena from this theory to disturbing causes. At the same time we confess that what we call disturbing causes are simply those parts of the true circumstances which we do not know or have neglected, and we endeavour in future to take account of them. We thus acknowledge that the so-called disturbance is a mere figment of the mind, not a fact of nature, and that in natural action there is no disturbance.

  But this is not the only way in which the harmony of the material with the mental operation may be disturbed. The mind of the mathematician is subject to many disturbing causes, such as fatigue, loss of memory, and hasty conclusions; and it is found that, from these and other causes, mathematicians make mistakes.

I am not prepared to deny that, to some mind of a higher order than ours, each of these errors might be traced to the regular operation of the laws of actual thinking; in fact we ourselves often do detect, not only errors of calculation, but the causes of these errors. This, however, by no means alters our conviction that they are errors, and that one process of thought is right and another process wrong. I

One of the most profound mathematicians and thinkers of our time, the late George Boole, when reflecting on the precise and almost mathematical character of the laws of right thinking as compared with the exceedingly perplexing though perhaps equally determinate laws of actual and fallible thinking, was led to another of those points of view from which Science seems to look out into a region beyond her own domain.

"We must admit," he says, "that there exist laws" (of thought) "which even the rigour of their mathematical forms does not preserve from violation. We must ascribe to them an authority, the essence of which does not consist in power, a supremacy which the analogy of the inviolable order of the natural world in no way assists us to comprehend."

***

Introductory Lecture on Experimental Physics.

James Clerk Maxwell

The University of Cambridge, in accordance with that law of its evolution, by which, while maintaining the strictest continuity between the successive phases of its history, it adapts itself with more or less promptness to the requirements of the times, has lately instituted a course of Experimental

Physics. This course of study, while it requires us to maintain in action all those powers of attention and analysis which have been so long cultivated in the University, calls on us to exercise our senses in observation, and our hands in manipulation. The familiar apparatus of pen, ink, and paper will no longer be sufficient for us, and we shall require more room than that afforded by a seat at a desk, and a wider area than that of the black board. We owe it to the munificence of our Chancellor, that, whatever be the character in other respects of the experiments which we hope hereafter to conduct, the material facilities for their full development will be upon a scale which has not hitherto been surpassed.

The main feature, therefore, of Experimental Physics at Cambridge is the Devonshire Physical Laboratory, and I think it desirable that on the present occasion, before we enter on the details of any special study, we should consider by what means we, the University of Cambridge, may, as a living body, appropriate and vitalise this new organ, the outward shell of which we expect soon to rise before us. The course of study at this University has always included Natural Philosophy, as well as Pure Mathematics. To diffuse a sound knowledge of Physics, and to imbue the minds of our students with correct dynamical principles, have been long regarded as among our highest functions, and very few of us can now place ourselves in the mental condition in which even such philosophers as the great Descartes were involved in the days before Newton had announced the true laws of the motion of bodies. Indeed the cultivation and diffusion of sound dynamical ideas has already effected a great change in the language and thoughts even of those who make no pretensions to science, and we are daily receiving fresh proofs that the popularisation of scientific doctrines is producing as great an alteration in the mental state of society as the material applications of science are effecting in its outward life. Such indeed is the respect paid to science, that the most absurd opinions may become current, provided they are expressed in language, the sound of which recals some well-known scientific phrase. If society is thus prepared to receive all kinds of scientific doctrines, it is our part to provide for the diffusion and cultivation, not only of true scientific principles, but of a spirit of sound criticism, founded on an examination of the evidences on which statements apparently scientific depend.

When we shall be able to employ in scientific education, not only the trained attention of the student, and his familiarity with symbols, but the keenness of his eye, the quickness of his ear, the delicacy of his touch, and the adroitness of his fingers, we shall not only extend our influence over a class of men who are not fond of cold abstractions, but, by opening at once all the gateways of knowledge, we shall ensure the association of the doctrines of science with those elementary sensations which form the obscure background of all our conscious thoughts, and which lend a vividness and relief to ideas, which, when presented as mere abstract terms, are apt to fade entirely from the memory.

In a course of Experimental Physics we may consider either the Physics or the Experiments as the leading feature. We may either employ the experiments to illustrate the phenomena of a particular branch of Physics, or we may make some physical research in order to exemplify a particular experimental method. In the order of time, we should begin, in the Lecture Room, with a course of lectures on some branch of Physics aided by experiments of illustration, and conclude, in the Laboratory, with a course of experiments of research.

Let me say a few words on these two classes of experiments,--Experiments of Illustration and Experiments of Research. The aim of an experiment of illustration is to throw light upon some scientific idea so that the student may be enabled to grasp it. The circumstances of the experiment are so arranged that the phenomenon which we wish to observe or to exhibit is brought into prominence, instead of being obscured and entangled among other phenomena, as it is when it occurs in the ordinary course of nature. To exhibit illustrative experiments, to encourage others to make them, and to cultivate in every way the ideas on which they throw light, forms an important part of our duty. The simpler the materials of an illustrative experiment, and the more familiar they are to the student, the more thoroughly is he likely to acquire the idea which it is meant to illustrate. The educational value of such experiments is often inversely proportional to the complexity of the apparatus. The student who uses home-made apparatus, which is always going wrong, often learns more than one who has the use of carefully adjusted instruments, to which he is apt to trust, and which he dares not take to pieces.

It is very necessary that those who are trying to learn from books the facts of physical science should be enabled by the help of a few illustrative experiments to recognise these facts when they meet with them out of doors. Science appears to us with a very different aspect after we have found out that it is not in lecture rooms only, and by means of the electric light projected on a screen, that we may witness physical phenomena, but that we may find illustrations of the highest doctrines of science in games and gymnastics, in travelling by land and by water, in storms of the air and of the sea, and wherever there is matter in motion.

This habit of recognising principles amid the endless variety of their action can never degrade our sense of the sublimity of nature, or mar our enjoyment of its beauty. On the contrary, it tends to rescue our scientific ideas from that vague condition in which we too often leave them, buried among the other products of a lazy credulity, and to raise them into their proper position among the doctrines in which our faith is so assured, that we are ready at all times to act on them.

Experiments of illustration may be of very different kinds. Some may be adaptations of the commonest operations of ordinary life, others may be carefully arranged exhibitions of some phenomenon which occurs only under peculiar conditions. They all, however, agree in this, that their aim is to present some phenomenon to the senses of the student in such a way that he may associate with it the appropriate scientific idea. When he has grasped this idea, the experiment which illustrates it has served its purpose.

In an experiment of research, on the other hand, this is not the principal aim. It is true that an experiment, in which the principal aim is to see what happens under certain conditions, may be regarded as an experiment of research by those who are not yet familiar with the result, but in experimental researches, strictly so called, the ultimate object is to measure something which we have already seen--to obtain a numerical estimate of some magnitude.

Experiments of this class--those in which measurement of some kind is involved, are the proper work of a Physical Laboratory. In every experiment we have first to make our senses familiar with the phenomenon, but we must not stop here, we must find out which of its features are capable of measurement, and what measurements are required in order to make a complete specification of the phenomenon. We must then make these measurements, and deduce from them the result which we require to find.

This characteristic of modern experiments--that they consist principally of measurements,--is so prominent, that the opinion seems to have got abroad, that in a few years all the great physical constants will have been approximately estimated, and that the only occupation which will then be left to men of science will be to carry on these measurements to another place of decimals.

If this is really the state of things to which we are approaching, our Laboratory may perhaps become celebrated as a place of conscientious labour and consummate skill, but it will be out of place in the University, and ought rather to be classed with the other great workshops of our country, where equal ability is directed to more useful ends.

But we have no right to think thus of the unsearchable riches of creation, or of the untried fertility of those fresh minds into which these riches will continue to be poured. It may possibly be true that, in some of those fields of discovery which lie open to such rough observations as can be made without artificial methods, the great explorers of former times have appropriated most of what is valuable, and that the gleanings which remain are sought after, rather for their abstruseness, than for their intrinsic worth. But the history of science shews that even during that phase of her progress in which she devotes herself to improving the accuracy of the numerical measurement of quantities with which she has long been familiar, she is preparing the materials for the subjugation of new regions, which would have remained unknown if she had been contented with the rough methods of her early pioneers. I might bring forward instances gathered from every branch of science, shewing how the labour of careful measurement has been rewarded by the discovery of new fields of research, and by the development of new scientific ideas. But the history of the science of terrestrial magnetism affords us a sufficient example of what may be done by Experiments in Concert, such as we hope some day to perform in our Laboratory.

That celebrated traveller, Humboldt, was profoundly impressed with the scientific value of a combined effort to be made by the observers of all nations, to obtain accurate measurements of the magnetism of the earth; and we owe it mainly to his enthusiasm for science, his great reputation and his wide-spread influence, that not only private men of science, but the governments of most of the civilised nations, our own among the number, were induced to take part in the enterprise. But the actual working out of the scheme, and the arrangements by which the labours of the observers were so directed as to obtain the best results, we owe to the great mathematician Gauss, working along with Weber, the future founder of the science of electro-magnetic measurement, in the magnetic observatory of Gottingen, and aided by the skill of the instrument-maker Leyser. These men, however, did not work alone. Numbers of scientific men joined the Magnetic Union, learned the use of the new instruments and the new methods of reducing the observations; and in every city of Europe you might see them, at certain stated times, sitting, each in his cold wooden shed, with his eye fixed at the telescope, his ear attentive to the clock, and his pencil recording in his note-book the instantaneous position of the suspended magnet.

Bacon's conception of "Experiments in concert" was thus realised, the scattered forces of science were converted into a regular army, and emulation and jealousy became out of place, for the results obtained by any one observer were of no value till they were combined with those of the others.

The increase in the accuracy and completeness of magnetic observations which was obtained by the new method, opened up fields of research which were hardly suspected to exist by those whose observations of the magnetic needle had been conducted in a more primitive manner. We must reserve for its proper place in our course any detailed description of the disturbances to which the magnetism of our planet is found to be subject. Some of these disturbances are periodic, following the regular courses of the sun and moon. Others are sudden, and are called magnetic storms, but, like the storms of the atmosphere, they have their known seasons of frequency. The last and the most mysterious of these magnetic changes is that secular variation by which the whole character of the earth, as a great magnet, is being slowly modified, while the magnetic poles creep on, from century to century, along their winding track in the polar regions.

We have thus learned that the interior of the earth is subject to the influences of the heavenly bodies, but that besides this there is a constantly progressive change going on, the cause of which is entirely unknown. In each of the magnetic observatories throughout the world an arrangement is at work, by means of which a suspended magnet directs a ray of light on a preparred sheet of paper moved by clockwork. On that paper the never-resting heart of the earth is now tracing, in telegraphic symbols which will one day be interpreted, a record of its pulsations and its flutterings, as well as of that slow but mighty working which warns us that we must not suppose that the inner history of our planet is ended.

But this great experimental research on Terrestrial Magnetism produced lasting effects on the progress of science in general. I need only mention one or two instances. The new methods of measuring forces were successfully applied by Weber to the numerical determination of all the phenomena of electricity, and very soon afterwards the electric telegraph, by conferring a commercial value on exact numerical measurements, contributed largely to the advancement, as well as to the diffusion of scientific knowledge.

But it is not in these more modern branches of science alone that this influence is felt. It is to Gauss, to the Magnetic Union, and to magnetic observers in general, that we owe our deliverance from that absurd method of estimating forces by a variable standard which prevailed so long even among men of science. It was Gauss who first based the practical measurement of magnetic force (and therefore of every other force) on those long established principles, which, though they are embodied in every dynamical equation, have been so generally set aside, that these very equations, though correctly given in our Cambridge textbooks, are usually explained there by assuming, in addition to the variable standard of force, a variable, and therefore illegal, standard of mass.

Such, then, were some of the scientific results which followed in this case from bringing together mathematical power, experimental sagacity, and manipulative skill, to direct and assist the labours of a body of zealous observers. If therefore we desire, for our own advantage and for the honour of our University, that the Devonshire Laboratory should be successful, we must endeavour to maintain it in living union with the other organs and faculties of our learned body. We shall therefore first consider the relation in which we stand to those mathematical studies which have so long flourished among us, which deal with our own subjects, and which differ from our experimental studies only in the mode in which they are presented to the mind.

There is no more powerful method for introducing knowledge into the mind than that of presenting it in as many different ways as we can. When the ideas, after entering through different gateways, effect a junction in the citadel of the mind, the position they occupy becomes impregnable. Opticians tell us that the mental combination of the views of an object which we obtain from stations no further apart than our two eyes is sufficient to produce in our minds an impression of the solidity of the object seen; and we find that this impression is produced even when we are aware that we are really looking at two flat pictures placed in a stereoscope. It is therefore natural to expect that the knowledge of physical science obtained by the combined use of mathematical analysis and experimental research will be of a more solid, available, and enduring kind than that possessed by the mere mathematician or the mere experimenter.

But what will be the effect on the University, if men Pursuing that course of reading which has produced so many distinguished Wranglers, turn aside to work experiments? Will not their attendance at the Laboratory count not merely as time withdrawn from their more legitimate studies, but as the introduction of a disturbing element, tainting their mathematical conceptions with material imagery, and sapping their faith in the formulae of the textbook? Besides this, we have already heard complaints of the undue extension of our studies, and of the strain put upon our questionists by the weight of learning which they try to carry with them into the Senate-House. If we now ask them to get up their subjects not only by books and writing, but at the same time by observation and manipulation, will they not break down altogether? The Physical Laboratory, we are told, may perhaps be useful to those who are going out in Natural Science, and who do not take in Mathematics, but to attempt to combine both kinds of study during the time of residence at the University is more than one mind can bear.

No doubt there is some reason for this feeling. Many of us have already overcome the initial difficulties of mathematical training. When we now go on with our study, we feel that it requires exertion and involves fatigue, but we are confident that if we only work hard our progress will be certain.

Some of us, on the other hand, may have had some experience of the routine of experimental work. As soon as we can read scales, observe times, focus telescopes, and so on, this kind of work ceases to require any great mental effort. We may perhaps tire our eyes and weary our backs, but we do not greatly fatigue our minds.

It is not till we attempt to bring the theoretical part of our training into contact with the practical that we begin to experience the full effect of what Faraday has called "mental inertia"--not only the difficulty of recognising, among the concrete objects before us, the abstract relation which we have learned from books, but the distracting pain of wrenching the mind away from the symbols to the objects, and from the objects back to the symbols. This however is the price we have to pay for new ideas.

But when we have overcome these difficulties, and successfully bridged over the gulph between the abstract and the concrete, it is not a mere piece of knowledge that we have obtained: we have acquired the rudiment of a permanent mental endowment. When, by a repetition of efforts of this kind, we have more fully developed the scientific faculty, the exercise of this faculty in detecting scientific principles in nature, and in directing practice by theory, is no longer irksome, but becomes an unfailing source of enjoyment, to which we return so often, that at last even our careless thoughts begin to run in a scientific channel.

I quite admit that our mental energy is limited in quantity, and I know that many zealous students try to do more than is good for them. But the question about the introduction of experimental study is not entirely one of quantity. It is to a great extent a question of distribution of energy. Some distributions of energy, we know, are more useful than others, because they are more available for those purposes which we desire to accomplish.

Now in the case of study, a great part of our fatigue often arises, not from those mental efforts by which we obtain the mastery of the subject, but from those which are spent in recalling our wandering thoughts; and these efforts of attention would be much less fatiguing if the disturbing force of mental distraction could be removed.

This is the reason why a man whose soul is in his work always makes more progress than one whose aim is something not immediately connected with his occupation. In the latter case the very motive of which he makes use to stimulate his flagging powers becomes the means of distracting his mind from the work before him.

There may be some mathematicians who pursue their studies entirely for their own sake. Most men, however, think that the chief use of mathematics is found in the interpretation of nature. Now a man who studies a piece of mathematics in order to understand some natural phenomenon which he has seen, or to calculate the best arrangement of some experiment which he means to make, is likely to meet with far less distraction of mind than if his sole aim had been to sharpen his mind for the successful practice of the Law, or to obtain a high place in the Mathematical Tripos.

I have known men, who when they were at school, never could see the good of mathematics, but who, when in after life they made this discovery, not only became eminent as scientific engineers, but made considerable progress in the study of abstract mathematics. If our experimental course should help any of you to see the good of mathematics, it will relieve us of much anxiety, for it will not only ensure the success of your future studies, but it will make it much less likely that they will prove injurious to your health.

But why should we labour to prove the advantage of practical science to the University? Let us rather speak of the help which the University may give to science, when men well trained in mathematics and enjoying the advantages of a well-appointed Laboratory, shall unite their efforts to carry out some experimental research which no solitary worker could attempt.

At first it is probable that our principal experimental work must be the illustration of particular branches of science, but as we go on we must add to this the study of scientific methods, the same method being sometimes illustrated by its application to researches belonging to different branches of science.

We might even imagine a course of experimental study the arrangement of which should be founded on a classification of methods, and not on that of the objects of investigation. A combination of the two plans seems to me better than either, and while we take every opportunity of studying methods, we shall take care not to dissociate the method from the scientific research to which it is applied, and to which it owes its value.

We shall therefore arrange our lectures according to the classification of the principal natural phenomena, such as heat, electricity, magnetism and so on.

In the laboratory, on the other hand, the place of the different instruments will be determined by a classification according to methods, such as weighing and measuring, observations of time, optical and electrical methods of observation, and so on.

The determination of the experiments to be performed at a particular time must often depend upon the means we have at command, and in the case of the more elaborate experiments, this may imply a long time of preparation, during which the instruments, the methods, and the observers themselves, are being gradually fitted for their work. When we have thus brought together the requisites, both material and intellectual, for a particular experiment, it may sometimes be desirable that before the instruments are dismounted and the observers dispersed, we should make some other experiment, requiring the same method, but dealing perhaps with an entirely different class of physical phenomena.

Our principal work, however, in the Laboratory must be to acquaint ourselves with all kinds of scientific methods, to compare them, and to estimate their value. It will, I think, be a result worthy of our University, and more likely to be accomplished here than in any private laboratory, if, by the free and full discussion of the relative value of different scientific procedures, we succeed in forming a school of scientific criticism, and in assisting the development of the doctrine of method.

But admitting that a practical acquaintance with the methods of Physical Science is an essential part of a mathematical and scientific education, we may be asked whether we are not attributing too much importance to science altogether as part of a liberal education.

Fortunately, there is no question here whether the University should continue to be a place of liberal education, or should devote itself to preparing young men for particular professions. Hence though some of us may, I hope, see reason to make the pursuit of science the main business of our lives, it must be one of our most constant aims to maintain a living connexion between our work and the other liberal studies of Cambridge, whether literary, philological, historical or philosophical.

There is a narrow professional spirit which may grow up among men of science, just as it does among men who practise any other special business. But surely a University is the very place where we should be able to overcome this tendency of men to become, as it were, granulated into small worlds, which are all the more worldly for their very smallness. We lose the advantage of having men of varied pursuits collected into one body, if we do not endeavour to imbibe some of the spirit even of those whose special branch of learning is different from our own.

It is not so long ago since any man who devoted himself to geometry, or to any science requiring continued application, was looked upon as necessarily a misanthrope, who must have abandoned all human interests, and betaken himself to abstractions so far removed from the world of life and action that he has become insensible alike to the attractions of pleasure and to the claims of duty.

In the present day, men of science are not looked upon with the same awe or with the same suspicion. They are supposed to be in league with the material spirit of the age, and to form a kind of advanced Radical party among men of learning.

We are not here to defend literary and historical studies. We admit that the proper study of mankind is man. But is the student of science to be withdrawn from the study of man, or cut off from every noble feeling, so long as he lives in intellectual fellowship with men who have devoted their lives to the discovery of truth, and the results of whose enquiries have impressed themselves on the ordinary speech and way of thinking of men who never heard their names? Or is the student of history and of man to omit from his consideration the history of the origin and diffusion of those ideas which have produced so great a difference between one age of the world and another?

It is true that the history of science is very different from the science of history. We are not studying or attempting to study the working of those blind forces which, we are told, are operating on crowds of obscure people, shaking principalities and powers, and compelling reasonable men to bring events to pass in an order laid down by philosophers.

The men whose names are found in the history of science are not mere hypothetical constituents of a crowd, to be reasoned upon only in masses. We recognise them as men like ourselves, and their actions and thoughts, being more free from the influence of passion, and recorded more accurately than those of other men, are all the better materials for the study of the calmer parts of human nature.

But the history of science is not restricted to the enumeration of successful investigations. It has to tell of unsuccessful inquiries, and to explain why some of the ablest men have failed to find the key of knowledge, and how the reputation of others has only given a firmer footing to the errors into which they fell.

The history of the development, whether normal or abnormal, of ideas is of all subjects that in which we, as thinking men, take the deepest interest. But when the action of the mind passes out of the intellectual stage, in which truth and error are the alternatives, into the more violently emotional states of anger and passion, malice and envy, fury and madness; the student of science, though he is obliged to recognise the powerful influence which these wild forces have exercised on mankind, is perhaps in some measure disqualified from pursuing the study of this part of human nature.

But then how few of us are capable of deriving profit from such studies. We cannot enter into full sympathy with these lower phases of our nature without losing some of that antipathy to them which is our surest safeguard against a reversion to a meaner type, and we gladly return to the company of those illustrious men who by aspiring to noble ends, whether intellectual or practical, have risen above the region of storms into a clearer atmosphere, where there is no misrepresentation of opinion, nor ambiguity of expression, but where one mind comes into closest contact with another at the point where both approach nearest to the truth.

I propose to lecture during this term on Heat, and, as our facilities for experimental work are not yet fully developed, I shall endeavour to place before you the relative position and scientific connexion of the different branches of the science, rather than to discuss the details of experimental methods.

We shall begin with Thermometry, or the registration of temperatures, and Calorimetry, or the measurement of quantities of heat. We shall then go on to Thermodynamics, which investigates the relations between the thermal properties of bodies and their other dynamical properties, in so far as these relations may be traced without any assumption as to the particular constitution of these bodies.

The principles of Thermodynamics throw great light on all the phenomena of nature, and it is probable that many valuable applications of these principles have yet to be made; but we shall have to point out the limits of this science, and to shew that many problems in nature, especially those in which the Dissipation of Energy comes into play, are not capable of solution by the principles of Thermodynamics alone, but that in order to understand them, we are obliged to form some more definite theory of the constitution of bodies.

Two theories of the constitution of bodies have struggled for victory with various fortunes since the earliest ages of speculation: one is the theory of a universal plenum, the other is that of atoms and void.

The theory of the plenum is associated with the doctrine of mathematical continuity, and its mathematical methods are those of the Differential Calculus, which is the appropriate expression of the relations of continuous quantity.

The theory of atoms and void leads us to attach more importance to the doctrines of integral numbers and definite proportions; but, in applying dynamical principles to the motion of immense numbers of atoms, the limitation of our faculties forces us to abandon the attempt to express the exact history of each atom, and to be content with estimating the average condition of a group of atoms large enough to be visible. This method of dealing with groups of atoms, which I may call the statistical method, and which in the present state of our knowledge is the only available method of studying the properties of real bodies, involves an abandonment of strict dynamical principles, and an adoption of the mathematical methods belonging to the theory of probability. It is probable that important results will be obtained by the application of this method, which is as yet little known and is not familiar to our minds. If the actual history of Science had been different, and if the scientific doctrines most familiar to us had been those which must be expressed in this way, it is possible that we might have considered the existence of a certain kind of contingency a self-evident truth, and treated the doctrine of philosophical necessity as a mere sophism.

About the beginning of this century, the properties of bodies were investigated by several distinguished French mathematicians on the hypothesis that they are systems of molecules in equilibrium. The somewhat unsatisfactory nature of the results of these investigations produced, especially in this country, a reaction in favour of the opposite method of treating bodies as if they were, so far at least as our experiments are concerned, truly continuous. This method, in the hands of Green, Stokes, and others, has led to results, the value of which does not at all depend on what theory we adopt as to the ultimate constitution of bodies.

One very important result of the investigation of the properties of bodies on the hypothesis that they are truly continuous is that it furnishes us with a test by which we can ascertain, by experiments on a real body, to what degree of tenuity it must be reduced before it begins to give evidence that its properties are no longer the same as those of the body in mass. Investigations of this kind, combined with a study of various phenomena of diffusion and of dissipation of energy, have recently added greatly to the evidence in favour of the hypothesis that bodies are systems of molecules in motion.

I hope to be able to lay before you in the course of the term some of the evidence for the existence of molecules, considered as individual bodies having definite properties. The molecule, as it is presented to the scientific imagination, is a very different body from any of those with which experience has hitherto made us acquainted.

In the first place its mass, and the other constants which define its properties, are absolutely invariable; the individual molecule can neither grow nor decay, but remains unchanged amid all the changes of the bodies of which it may form a constituent.

In the second place it is not the only molecule of its kind, for there are innumerable other molecules, whose constants are not approximately, but absolutely identical with those of the first molecule, and this whether they are found on the earth, in the sun, or in the fixed stars.

By what process of evolution the philosophers of the future will attempt to account for this identity in the properties of such a multitude of bodies, each of them unchangeable in magnitude, and some of them separated from others by distances which Astronomy attempts in vain to measure, I cannot conjecture. My mind is limited in its power of speculation, and I am forced to believe that these molecules must have been made as they are from the beginning of their existence.

I also conclude that since none of the processes of nature, during their varied action on different individual molecules, have produced, in the course of ages, the slightest difference between the properties of one molecule and those of another, the history of whose combinations has been different, we cannot ascribe either their existence or the identity of their properties to the operation of any of those causes which we call natural.

Is it true then that our scientific speculations have really penetrated beneath the visible appearance of things, which seem to be subject to generation and corruption, and reached the entrance of that world of order and perfection, which continues this day as it was created, perfect in number and measure and weight?

We may be mistaken. No one has as yet seen or handled an individual molecule, and our molecular hypothesis may, in its turn, be supplanted by some new theory of the constitution of matter; but the idea of the existence of unnumbered individual things, all alike and all unchangeable, is one which cannot enter the human mind and remain without fruit.

But what if these molecules, indestructible as they are, turn out to be not substances themselves, but mere affections of some other substance?

According to Sir W. Thomson's theory of Vortex Atoms, the substance of which the molecule consists is a uniformly dense plenum, the properties of which are those of a perfect fluid, the molecule itself being nothing but a certain motion impressed on a portion of this fluid, and this motion is shewn, by a theorem due to Helmholtz, to be as indestructible as we believe a portion of matter to be.

If a theory of this kind is true, or even if it is conceivable, our idea of matter may have been introduced into our minds through our experience of those systems of vortices which we call bodies, but which are not substances, but motions of a substance; and yet the idea which we have thus acquired of matter, as a substance possessing inertia, may be truly applicable to that fluid of which the vortices are the motion, but of whose existence, apart from the vortical motion of some of its parts, our experience gives us no evidence whatever.

It has been asserted that metaphysical speculation is a thing of the past, and that physical science has extirpated it. The discussion of the categories of existence, however, does not appear to be in danger of coming to an end in our time, and the exercise of speculation continues as fascinating to every fresh mind as it was in the days of Thales.

###

www.ingramcontent.com/pod-product-compliance
Lightning Source LLC
Chambersburg PA
CBHW070923180526
45168CB00005B/2135